I0469316

Federal Emergency Management Agency

United States Fire Administration

National Fire Data Center

ESTABLISHING A RELATIONSHIP BETWEEN

ALCOHOL AND CASUALTIES OF FIRE

Prepared by:

TriData Corporation
1000 Wilson Boulevard
Arlington, Virginia 22209

October 1999

This publication was produced under contract by TriData Corporation for the United States Fire Administration, Federal Emergency Management Agency. Any information, findings, conclusions, or recommendations expressed in this publication do not necessarily reflect the views of the Federal Emergency Management Agency or the United States Fire Administration.

TABLE OF CONTENTS

I. EXECUTIVE SUMMARY

Though the rate has significantly decreased, the United States continued into the late 90's with one of the highest fire death rates in the industrialized world. Given the advancements in fire prevention, including public education, building design, consumer product safety, and sophisticated levels of the fire protection in this country, it is puzzling to many as to why this is so. In an effort to identify the underlying problem(s), researchers have been delving deeper into the extent to which human behavior affects our fire losses.

The connection between alcohol and the ignition, detection, and escape from the fire has been broadly examined by numerous medical and fire protection organization studies. A series of landmark studies undertaken by the Johns Hopkins University and the National Bureau of Standards in the 1970's were among the first to discover a definitive link between alcohol consumption and fire deaths. Many studies have now confirmed their general findings.

Alcohol intoxication may increase the risk of initiating a fire by impairing one's judgment and coordination. An intoxicated individual who is smoking may also succumb to the depressant effects of alcohol, fall asleep and drop a lit cigarette on upholstery or clothing. Intoxication also acutely diminishes one's ability to detect a fire. Under the sedative effects of alcohol, an alcohol-impaired person may fail to notice the smell of smoke, or fail to hear a smoke alarm. Escape from a fire can be hampered by the loss of motor coordination and mental clarity caused by alcohol, even when warning signs are heeded. Furthermore, burns are more physiologically damaging in the presence of alcohol.

Several researchers have found that about half of all adult fire fatalities were under the influence of alcohol at the time of the fire. Men have been found to consistently outnumber women among fire casualties and do so with even greater disparity for fire victims under the influence of alcohol. In addition, the younger adult population (ages 15 to 34) seems to incur the greatest number of alcohol-impaired fire casualties. Drinking behaviors that are characteristic of various age groups and sexes may explain these findings.

Studies have also provided conclusive evidence supporting the deleterious effects of chronic and acute alcohol abuse on the occurrence and recovery from burn injuries. Burn injury victims have been found to be disproportionately likely to have been intoxicated at the time of injury or known to be heavy drinkers. From a physiological standpoint, burn victims with histories of alcoholism tend to have longer hospital stays, more complications, and higher mortality rates as a result of their burns.

Questions still remain as to the extent that alcohol affects fire losses. How do we explain the fact that some industrialized countries with some of the highest alcohol consumption rates per capita, e.g. Germany and the Netherlands, have relatively low fire death rates? Researchers have suggested that alcohol-related unintentional injuries have more to do with alcohol drinking patterns than the total amount of alcohol consumed per capita. Who drinks, where they drink, what they drink, and under what social, cultural, and religious circumstances they drink are perhaps more significant factors than the amount of alcohol consumed. A lone drinker at home is probably at greater risk of a fire emergency than a group of people drinking at a bar or restaurant. Moreover, the number of drinks consumed in a single sitting seems to matter a great deal.

Alcoholics have a disproportionately high rate of fire fatalities relative to their percentage of the total population. Non-intoxicated fire victims also may be affected by alcohol: they may have been entrusted to the care of an alcohol-impaired individual. These fire fatalities would not be reported as related to alcohol when blood alcohol levels (BALs) are taken of victims only. As a result, the estimated number of alcohol-related fire casualties as well as the magnitude of the problem may be underestimated.

Smoking fires are the leading cause of fire fatalities. The incidence of such fatal fires is higher among those who are under the influence of alcohol and most smoking-related fire fatalities have some connection to alcohol consumption.

In summary, there is a clear connection of alcohol and fire fatalities. Unlike the connection between alcohol consumption and vehicle fatalities, the connection is not often referred to in prevention programs, nor has much been done to address the problem.

II. INTRODUCTION

Alcohol consumption plays a key role in American society. It is heavily embedded in our social framework and tied to many of our traditions. Every socio-economic class and race consumes alcohol. It is heavily advertised, readily available and is produced in a multitude of forms at a variety of costs. Yet, the negative impact of alcohol on the nation may be more devastating than any other substance. It is centrally related to a number of persistent threats to public health. These include unintentional injuries such as burns, falls, drownings, and motor vehicle crashes; violent crimes such as rape, homicide, and arson; chronic diseases such as cancer, enteric disease, and cardiovascular disease; and social problems such as unintended pregnancies, the promulgation of sexually transmitted diseases, child neglect, "failed" marriages, and homelessness.

Numerous studies have shown the deleterious effects of alcohol on cognitive and physical functions.[1,2] Alcohol affects judgment, balance and motor coordination, all of which lead to an increased rate of unintentional injury.[3] Many of the studies produced on the relationship between injury and alcohol impairment focus on injuries that result from motor vehicle crashes. Motor vehicle crashes are the leading cause of fatal injuries in the United States.[4] Even now, alcohol has been found to be involved in 40 percent of all motor vehicle fatalities.[5] This reality drives nationwide research and prevention efforts focusing on the role alcohol plays in these events.

---

[1] P.F. Waller, J.R. Stewart, A.R. Hansen, J.C. Stutts, C.L. Popkin and E.A. Rodgman. "The Potentiating Effects of Alcohol on Driver Injury." *Journal of the American Medical Association.* 1986. Sep 19;256(11):1461-6

[2] C.J. Cherpitel. "The Epidemiology of Alcohol Related Trauma." *Alcohol, Health and Research World.* 1992. 16(3):191-196.

[3] P.F. Waller, J.R. Stewart, A.R. Hansen, J.C. Stutts, C.L. Popkin and E.A. Rodgman. "The Potentiating Effects of Alcohol on Driver Injury." *Journal of the American Medical Association.* 1986. Sep 19;256(11):1461-6

[4] Centers for Disease Control and Prevention. "Report of Final Mortality Statistics – 1995." *Monthly Vital Statistics Report.* 1997. Jun 12:45(11)S2:1-33.

[5] "Traffic Safety Facts 1995: Alcohol," National Highway Traffic Safety Administration, National Center for Statistics & Analysis, 1996.

Fire, the fifth leading cause of unintentional deaths in the United States, causes between 4,000 and 5,000 lives and billions in property damage each year.[6] It is estimated that, like motor vehicle crashes, alcohol is involved in over 40 percent of all residential fire deaths.[7] Yet there is not a national initiative to address this problem.

It, too, is estimated that alcohol is involved in 40 percent of all residential fire deaths.[8] There is, however, no equivalent and comprehensive body of research for the relationship between alcohol and fire. A myriad of research exists when relating alcohol use and vehicular accidents, homicide, and suicide, but not for fires. For this reason, it may appear that alcohol is less related to the incidence of fire when compared to other unintentional injuries than it really is.

This lack of a national prevention and education campaign may be related to several significant obstacles hindering researchers' attempts to quantify the relationship between fire and alcohol use on a large scale. There is no nationwide attempt to collect baseline epidemiological data concerning pre-existing medical conditions or impairments present in the fatal fire casualty. Such an endeavor would demand closer collaboration between fire departments, fire investigators, pathologists, toxicologists and statisticians than current convention allows. Spot studies with a wide variety of time frames, sample sizes, and lack of controls have been the hallmark of studies examining alcohol and fires, not a sustained national data collection as exists for vehicle accidents.

Alcohol impairment has been one of the data items requested by the National Fire Incident Reporting System (NFIRS) in reporting the condition of a fire victim, but this information is not routinely reported by participating fire departments. Fire departments and investigators tend to be reluctant to record alcohol abuse by those who cause or are injured by fires. The reasoning behind their reluctance has ranged from a humanitarian concern for the victims of fire and their loved ones already suffering from the effects of a fire to a lack of training and authorization to test people for alcohol or drugs. In addition, there is the potential for legal ramifications concerning wrongful accusations about intoxication.

---

[6] U.S. Fire Administration, *Fire in the United States: 1987-1996*, Eleventh Edition, FA-173/August 1999, (Emmitsburg, MD: U.S. Fire Administration).
[7] S.W. Marshall, C.W. Runyan, *et al.* "Fatal Residential Fire: Who Dies and Who Survives?" The Journal of the American Medical Association.279(20):1633-1637, May 1998.

Data from independent studies have proven difficult to compare due to the variation in the definition of alcohol-impairment and in measurements made of BALs. [9] In other cases, the sample size is simply too small to derive any statistical significance in itself, but is still suggestive of a pattern. Without full cooperation and standardized techniques for data collection, it is difficult to perform an in-depth and integrated study into the involvement of alcohol and fires.

Much of what we now know about alcohol's involvement in fires has stemmed from medical literature and research rather than fire research, with two large exceptions: The Johns Hopkins and Minnesota studies, which will be discussed in the body of the report. The results of such studies are limited and precarious for a number of reasons. Medical studies have traditionally been sporadic and usually confined to a city's or state's medical records. In addition, not all burn centers routinely screen their burn or fire-related injury patients for alcohol use, so the number of missed patients varies greatly.

Medical scientists have addressed the issue of alcohol's relation to injuries caused by fires in many small-scale studies. By examining data from both burn centers and emergency departments a clear pattern has emerged. Alcohol use is associated with injury by fire. Unfortunately, without the integration of on-scene data, specifically the origins of particular fires, the activities of the injured at the time of the fire and their proximity to the point of ignition, the medical studies paint only half of a picture. There is no solid data on the mechanism by which alcohol is associated with fire injury.

The number of studies that examine the relationship between alcohol use and fires is increasing. Evidence exists that portrays alcohol's detrimental effects on fire losses and both human and property terms. People who use and abuse alcohol are being identified as a growing high-risk fire group that must be targeted by prevention efforts.

This report examines fire and alcohol studies performed by medical scientists, fire investigators, and social theorists. The physiological effects of long and short-term alcohol use are described. Alcohol's effect on behavior and its role in unintentional injuries is examined. The demographics of alcohol use and abuse as well as societal

---

[8] Ibid.

factors influencing alcohol use are noted. Caretakers who are under the influence of alcohol are also studied in regards to how their impairment may affect a dependent individual.

The demographics of fire and fire fatalities are described. The leading causes of fire fatalities are also discussed, specifically how and to what extent they are affected by alcohol. The influence of alcohol in these fatal fire events is examined.

It is the intent of this report to:

- Identify patterns and trends in alcohol use and fire casualties.

- Summarize the existing quantitative data on the involvement of alcohol in fire deaths.

- Summarize the impact of socio-economic and demographic variables on alcohol use and fire casualties.

- Identify those who are most affected by alcohol's involvement in fires.

- Analyze two case studies from Ontario and Minnesota to provide further input concerning the question of alcohol use and fire.

- Demonstrate the need for public awareness and education on the role of alcohol use in fatal fires.

---

[9] Blood alcohol level, also known as blood alcohol concentration, is the amount of alcohol in the blood. It is usually expressed in g/dl. The National Highway Transportation Safety Administration considers fatal vehicle crashes as alcohol-related if the individuals involved have BALs greater than .01g/dl.

III. ALCOHOL'S PHYSIOLOGIC EFFECTS

Alcohol affects the incidence and severity of unintentional injuries in a variety of ways. Alcohol depresses the central nervous system, potentially to the point of stupor, coma, and death. Individuals who consume large quantities of alcohol experience disordered thought patterns, impaired judgment, impaired perception, and a decrease in generalized motor control. As blood alcohol concentration increases in the body, alcohol's depressive effects are more prominent. The effects of alcohol are proportionately related to the concentration of alcohol in the blood, which vary from person to person depending on numerous individual characteristics.

Ethanol (ethyl alcohol) or "grain alcohol" is found in distilled spirits, wine, and beer. It is the agent responsible for the physiologic effects of alcoholic beverages. Ethanol is the most commonly used mood-altering drug in the United States.

When ingested, ethanol is absorbed into the blood stream unaltered in the stomach and upper digestive system. Fatty foods and/or milk can slow this absorption. As it is absorbed, ethanol is distributed to all bodily tissues and fluids in the same concentration as it is in blood. For this reason blood ethanol level tests (reported in mg/dl) are used to medically define the degree of intoxication. The drinker's size, sex, body build, and metabolism all influence blood alcohol concentration and subsequently the degree of intoxication. In time, ninety percent of the ingested ethanol reaches the liver where it is metabolized. The remaining ten percent is excreted unchanged in breath, perspiration, and urine.[10]

In the central nervous system, alcohol acts directly as a depressant. At low levels it results in motor coordination dysfunction and intellectual confusion. As judgment becomes impaired, the user may appear giddy, act excited, or become aggressive. This is a result of depression of inhibitions and impaired reasoning ability. The motor and intellectual effects become more pronounced as the level of alcohol in the blood stream is increased.[11]

---

[10] K.L. McCance and S.E. Huether. *Pathophysiology: The Biologic Basis for Disease in Adults and Children.* 1994. St. Louis, MO: Mosby. Pp 62-3.
[11] K.L. McCance and S.E. Huether. *Pathophysiology: The Biologic Basis for Disease in Adults and Children.* 1994. St. Louis, MO: Mosby. Pp 62-3.

At higher blood concentration levels, respiration becomes affected. If these effects become pronounced due to increasing blood ethanol concentrations (at about 400 mg/dl or 0.40 percent BAL) or persist for a long period, the user can become comatose or die.[12]

Chronic alcohol use is responsible for pathophysiologic changes in almost every tissue in the body. Chronic alcohol use is related to an increased tendency toward hypertension, higher incidence of acute and chronic pancreatitis, regressive changes in skeletal muscle, and cirrhosis.[13] Research shows that chronic alcohol users have significantly shortened lifespans.[14]

---

[12] K.L. McCance and S.E. Huether. *Pathophysiology: The Biologic Basis for Disease in Adults and Children*. 1994. St. Louis, MO: Mosby. Pp 62-3.
[13] K.L. McCance and S.E. Huether. *Pathophysiology: The Biologic Basis for Disease in Adults and Children*. 1994. St. Louis, MO: Mosby. Pp 62-3.
[14] R.M. Costello, P. Parsons-Manders and S.L. Schneider. "Alcoholic Mortality: A Twelve Year Follow-up." *American Journal of Drug and Alcohol Abuse*. 1978. 5(2):199-210.

IV. ALCOHOL'S INVOLVEMENT IN UNINTENTIONAL INJURIES AND FIRE CASUALTIES

Alcohol and Unintentional Injuries

Unintentional injuries account for over 90,000 deaths and millions of non-fatal injuries each year.[15] They are also the leading cause of death for all people under the age of 44. Motor vehicle crashes, fires, burns, falls, drowning, and poisonings are the top ranking causes of death from unintentional injuries. Numerous studies have identified a significant relationship between alcohol and unintentional injuries and deaths. In fact, in emergency room and trauma center studies of alcohol and injury, alcohol was found to be associated with trauma cases rather than medical problems among a representative sample of patients admitted to the same facility.[16] That is, when a patient was admitted with an injury, the injury was likely to be alcohol-related. When a patient was admitted with a non-injury (e.g. illness), alcohol had little or no relationship to the illness. It has been estimated that nearly 50 percent of adult trauma patients in the United States are injured under the influence of alcohol.[17]

Short-term physiological effects of alcohol have been shown to diminish motor coordination and balance, as well as impair perception and judgment. The propensity for injuring oneself under these conditions is great. In a study performed by researchers from the Alcohol Research Center at the California Pacific Center Research Institute, the association between alcohol and injuries was examined among an emergency room patient sample. Of those identified with a positive BAL, 23 percent reported feeling drunk at the time of the incident, and of these, 45 percent felt the event could have been avoided had they not been drinking.[18]

Not only does alcohol impair physical capabilities; it has been theorized that alcohol perpetuates accident-prone behavior. A classic example of this is the drunk

---

[15] U.S. Department of Commerce, Bureau of the Census. *The Statistical Abstract of the United States – 1993 113th ed.* 1993. Washington, DC:GPO.

[16] C.J. Cherpitel, "Alcohol and Injuries: A Review of International Emergency Room Studies," *Addiction.*88:923-937, 1993.

[17] C.W. Dunn, D.M. Donovan , L.M. Gentilello, "Practical Guidelines for Performing Alcohol Interventions in Trauma Centers," *Journal of Trauma.* 42(2):299-304, February 1997.

driver who fails to fasten his seat belt and as a result of his alcohol-impaired driving ability, sustains life-threatening or fatal injuries. Researchers from the Boston University School of Public Health found that an average daily intake of 5 or more alcoholic beverages elevated the relative risk for injuries among a sample population. These drinkers were 1.7 times more likely to sustain an unintentional injury and twice as likely to require hospitalization as a result of their injuries when compared to non-drinkers.[19]

Alcohol and Fire Casualties

In the United States, fire-related injuries rank fifth among unintentional injuries, after motor vehicle crashes, poisoning, falls, and drowning.[20] The linking of alcohol abuse to fire casualties is a subset of the broader correlation between alcohol and traumatic accidents.

Some of the most persuasive data come from studies of traumatic injuries suffered by chronic alcohol abusers. Researchers in Toronto tracked the mortality experience of alcoholics and found their fire death risk to be 9.7 times greater than that of the rest of the population.[21]

> **Definitions**:
> *Acute Alcohol Abuse* – A single or specific situation of alcohol intoxication.
> *Chronic Alcohol Abuse* – The use of alcohol, for the purpose of intoxication, more often than once per week for a period exceeding six months.
> *Alcoholism* – An incurable disease defined in the Diagnostic Service Manual of the American Psychological Association. An alcoholic does not necessarily currently abuse alcohol.

In 1987, two researchers from the Boston University Medical School's Public Health Department conducted a pioneer study evaluating alcohol and fire victims. Jonathan Howland and Ralph Hingson performed a literature survey on alcohol involvement in fire fatalities and serious burns in the United States and abroad for the time period of 1958 to 1981, and looked for a pattern across various studies. The two researchers concluded that of the fatalities screened for BALs, a significant portion of the

---

[18] C.J. Cherpitel, "Injury and the Role of Alcohol: Countywide Emergency Room Data." *Alcoholism, Clinical & Experimental Research*. 18(3):679-84, June 1994.
[19] R.W. Hingson, R.I. Lederman and D.C. Walsh. "Employee Drinking Patterns and Accidental Injury: A Study of Four New England States." *Journal of Studies on Alcohol*. 46(4):298-303, July 1985.
[20] Centers for Disease Control and Prevention. Injury Mortality 1986-1992. Hyattsville, MD: Centers for Disease Control and Prevention, 1995.
[21] Schmidt and De Lint, "Causes of Death of Alcoholics," *Quarterly Journal of Studies on Alcohol* 33(1): pp. 171-85, 1972.

fire fatalities were legally drunk at 0.08 percent BAL or 0.1 percent BAL, depending on jurisdictional standards, or had elevated BALs. Half of all people over the age of 15 who died in fires had elevated BALs.[22] An earlier study that examined fire deaths in Memphis, Tennessee found that 83 percent of people between the age of 16 and 60 had been drinking at the time of their death.[23] While these data are relatively dated, all indications are that it remains accurate. A recently published study, which analyzed all fatal residential fires in North Carolina, revealed that for fire fatalities age 18 and older, 53 percent had BALs equivalent to over 0.1 percent and that 28 percent had histories of alcoholism.[24]

Alcohol and Burns

Alcohol use by the burn patient is a very specific concern. Chronic alcohol use has been shown to disrupt the immune system response to a significant burn.[25] This appears to be the result of alcohol-related damage to the spleen and lymph nodes hindering synthesis of a key immune system component.[26] Acute alcohol intoxication also poses a problem for the burn patient. Both burn injuries and alcohol (ethanol) intoxication suppress the immune system. When they occur together, a synergistic effect has been documented, resulting in a very suppressed immune response.[27, 28] In either case, chronic or acute, the presence of ethanol use in the burn patient results in a higher rate of infectious complications and increased mortality. Researchers from the University of Washington Burn Center in Seattle found alcoholic patients with burn injuries had an overall mortality rate three times that of non-alcoholics and also died of smaller burn

---

[22] J. Howland and R. Hingson, "Alcohol as a Risk factor for Injuries or Death Due to Fires and Burns: Review of the Literature." *Public Health Reports*.102(5):475-83, Sept-Oct. 1987.

[23] W.S. Hollis, "Drinking: Its Part in Fire Deaths," *Fire Journal*, pp. 10-12, May 1973.

[24] S.W. Marshall, C. W. Runyan, *et al.* "Fatal Residential Fire: Who Dies and Who Survives?" *The Journal of the American Medical Association*. 279(20):1633-1637, May 1998.

[25] M. Kawakami, A.A. Meyer, M.C. Johnson, S. DeSerres and H.D. Peterson. "Chronic Ethanol Exposure Before Injury Produces Greater Immune Dysfunction After Thermal Injury in Rats." *Journal of Trauma*. 1990. Jan;30(1):27-31.

[26] T. Tabata and A.A. Meyer. "Immunoglobulin M Synthesis After Burn Injury: The Effects of Chronic Ethanol on Postinjury Synthesis." *Journal of Burn Care and Rehabilitation*. 1995. Jul-Aug;16(4):400-6.

[27] M. Kawakami, B.R. Switzer, S.R. Herzog and A.A. Meyer. "Immune Suppression After Acute Ethanol Ingestion and Thermal Injury." *Journal of Surgical Research*. 1991. Sep;51(3):210-5

[28] Chronic Ethanol Intake and Burn Injury: Evidence for Synergistic Alteration in Gut and Immune Integrity," *Journal of Trauma*. 38(2): 198-207, February 1995.

wounds. Surviving alcoholic burn patients required significantly more intravenous antibiotics and fluids and longer hospitalization.[29]

Research examining the effect of alcohol on burns has shown that alcohol acts at the cellular level by preventing post-injury homeostasis. It also acts systemically by interfering with the body's vasoconstriction response to shock, a common complication of serious burn injuries. Two studies performed by the Department of Plastic and Reconstructive Surgery and Burn Unit in Cologne, Germany recently examined the relationship between alcohol use and burn injuries. Patients with positive BALs were found to have a significantly higher fatality rate (31 percent) in comparison to patients with a negative BAL (18 percent), even though both groups had equal total burn surface area and were of similar ages.[30] In a study of burn patients from the University of Florida's College of Medicine, researchers reported that the average length of stay for the alcohol group was 9 days longer than the average stay in the burn center, resulting in additional medical costs of $337,500.[31]

Some studies have found that alcoholics also comprise a large portion of the burn-injured population. A study performed in New York City found that 19 percent of burn victims treated at the New York Hospital Burn Center had an alcohol abuse problem, as detected from patient interviews using the standardized CAGE questionnaire on alcohol use.[32] While attempting to determine predictors for post-traumatic stress after burn injuries, researchers from Cornell University and Columbia University College of Physicians and Surgeons in New York City discovered that 24 percent of the patients screened had histories of alcoholism, and 18 percent had drug abuse histories. It is significant to note that only 7 percent of the adult population in general are alcoholics. The high percentage of alcoholics and other alcohol abusers among the burn population surveyed in these studies is suggestive of drinking patterns witnessed in alcoholics that elevates their risk for fire injury.

---

[29] J.D. Jones, B. Barber and L. Engrav, "Alcohol Use and Burn Injury." *Journal of Burn Care and Rehabilitation.* 12(2):148-152, March-April 1991.

[30] A. Haum, W. Perbix and H. J. Hack, "Alcohol and Drug Abuse in Burn Injuries" *Burns.* 21 (3): 194-1999. May 1995.

[31] P.S. Powers, B. Stevens, *et al.*, "Alcohol Disorders Among Patients with Burns: Crisis and Opportunity." *Journal of Burn Care and Rehabilitation.* 15(4):386-91, July-August 1994.

[32] L. Bernstein *et al.*, "Detection of Alcoholism Among Burn Patients," *Hospital Community Psychiatry.* 40: 255-6, 1992.

Alcohol and Smoke Inhalation

The overwhelming majority of fire fatalities perish as a result of smoke and toxic fume inhalation as opposed to burn injuries.[33, 34] Sensitivity to smoke is a key means by which the human body detects a fire. Mitchell *et al.* found evidence to suggest that alcohol not only impedes the detection of smoke, but also helps to facilitate its passage into the body.[35] In a study performed by researchers from the Queens Medical Center in England, the effect of ethyl alcohol ingestion on upper airway reflexes, e.g. coughing and sneezing, was examined for a sample of healthy, young men. This research found that ethyl alcohol caused depression of the upper airway reflexes, particularly when BALs were in excess of 0.1 percent BAL.[36] This finding supports the earlier research presented by Mitchell *et al.*, which revealed that in addition to enhancing the toxic effects of gases such as carbon monoxide, BALs depressed the cough reflex as well.

---

[33] H. Gormsen, N. Jeppesen and A. Lund, "The Causes of Death in Fire Victims," *Forensic Science International*. 24(2):107-11, February 1984.

[34] J.R. Hall and B. Harwood, "Smoke or Burns–Which is Deadlier?" *Fire Journal*. 89(1):38-43, January-February 1995.

[35] Mitchell *et al.*, "Effects of Ethanol and Carbon Monoxide upon Two Measures of Behavioral Incapacitation's on Rats," *Proceedings of the Western Pharmacology Society*, Sec.21, pp427-431(1978)

[36] R. Erskine, P. Murphy and J.A. Langton, "The Effect of Ethyl Alcohol on the Sensitivity of Upper Airway Reflexes." *Alcohol & Alcoholism*. 29(4):425-31, July 1994.

V. ALCOHOL, UNINTENTIONAL INJURIES, AND FIRE CASUALTIES

Alcohol and Alcohol's Role in Injury

Drinking Patterns

In 1995, a national survey of adults over the age of 18 in the United States was conducted to identify the drinking behaviors of Americans. In this survey performed by the Centers for Disease Control and Prevention (CDC), Americans over the age of 18 were asked if they drank at least 1 alcoholic beverage during the past month. The majority of adults (53 percent) said yes. Those individuals aged 18 to 24 and 25 to 34 accounted for the highest percentage of drinkers at 57 percent and 63 percent respectively. Of those aged 18 to 24, 71.8 percent reported drinking anywhere from 1 to 5 days per month. For those aged 25 to 34, 68 percent claimed the same frequency. In addition, respondents were asked how many drinks were consumed on the average for the days they drank. Respondents aged 18 to 24 comprised the largest percentage of the 5 drinks or more category, followed closely by the age bracket 25 to 34. In fact, the 5 drinks or more category constituted one-quarter of respondents age 18 to 24.[37]

The same survey recorded different, yet intriguing, results for respondents aged 65 and older. Approximately 35 percent reported drinking during the previous month. Of this group, the majority reported drinking only between 1 and 5 days per month. However, another one-quarter of respondents aged 65 and older claimed that they consumed alcohol anywhere from 21 to 31 days per month. The overwhelming majority of this same age group (63 percent) claimed that they consumed 1 to 2 alcoholic drinks on those days when they drank.

What does this mean? The youngest portion of the adult population surveyed, ages 18 to 24, is drinking with less frequency than the older groups, but with more intensity when they do drink. This phenomenon, dubbed binge drinking by researchers, is seen in high-school adolescents and young college adults. For the elderly community, the

---

[37] Behavioral Risk Factor Surveillance System, "Nationwide Alcohol Consumption," Online Prevalence Data, Centers for Disease Control and Prevention, 1995.

results are just the opposite; the elderly average more alcohol consuming days per month, but with less volume per session.

Alcohol's Role in Unintentional Injury

Drinking has been associated with increased risk for unintentional injuries of all types. Sensation seeking and risk-taking are prevalent among the younger populations that use or abuse alcohol. It is this type of behavior that is predictive of incurring unintentional injuries and accident statistics are a prime indicator of this. In a review of adolescent injuries and the involvement of alcohol, researchers from the Harborview Injury Prevention and Research Center in Seattle Washington reported that of 319 young adults aged 18 to 20 admitted to the trauma center, 41 percent tested positive for alcohol.[38] Approximately 22 percent were legally intoxicated at the time of their injury, indicated by BALs of 0.1 percent or more.

In 1996, researchers from the Edwin Albano Institute of Forensic Science performed a study aimed at quantifying the role of substance abuse in fatal fires occurring in New Jersey over a 7-year period. Records of fire fatalities of all ages were retrospectively examined for the nearly 30 percent of fire victims who tested positive for BALs of any concentration. Of these victims, 58 percent were between the ages of 21 and 50. That very same year, the National Center for Health Statistics reported that only 37 percent of all fire fatalities were age 21 to 50.

The very young and the very old have a much higher fire death rate than the average population. The risk of fire death drops considerably after the age of 5 and experiences minimal change until the age of 55, at which point the risk begins to rise substantially.[39] Young children and older adults are typically high-risk groups because they are physically unable to escape a fire. People between the ages of 5 and 55 years generally do not suffer from physical inability and are capable of removing themselves from danger, which is one explanation for their lower fire death rates. Alcohol-impaired fire fatalities exhibit an age pattern that is quite the opposite of overall fire death profile. The majority of alcohol-impaired deaths are between the ages of 18 and 40, the

---

[38] F.P. Rivara, J.G. Gurney, R.K. Ries and D.A. Seguin, "A Descriptive Study of Trauma, Alcohol, and Alcoholism in young adults." *Journal of Adolescent Health.* 13(8):663-667, December 1992.
[39] U.S. Fire Administration, *Fire in the United States: 1987-1996,* Eleventh Edition, FA-173/August 1999, (Emmitsburg, MD: U.S. Fire Administration).

population most expected to escape injury. This pattern suggests that a relationship exists between intoxication and the risk of injury from a fire.

Caretakers Under the Influence

When examining the prevalence of alcohol involvement in all types of unintentional injuries, children are not typically pictured. Alcohol impairment is viewed as an adult activity that precludes children by virtue of their age and innocence. Unfortunately, children are not exempt from the harmful effects of alcohol, primarily through no fault of their own. As an example, a recent study performed by the CDC found that for 60 percent of child passenger fatalities, the driver of the child was impaired.[40]

Numerous studies have examined the impact of parental drinking patterns on the health and well-being of children. Researchers from the Department of Pediatrics at the Albert Einstein College of Medicine found that the children of mothers characterized as problem drinkers had over twice the risk of serious injury as children of mothers who were non-drinkers.[41] They also concluded that children with two parents who are problem drinkers are at an even greater risk.

As previously noted, children are one of the highest risk groups for residential fire deaths. While not traditionally considered in studies examining alcohol and fire deaths, alcohol-related fires often affect them too. A study performed by Marshall *et al.* found that of the juvenile fatalities examined, approximately 15 percent died in fires where the surviving adult was impaired by alcohol or other drugs.[42] In another study performed by Hollis *et al.*, the authors stated that in a review of case files of fire fatalities under age 16 or over age 60, "case after case revealed that fire deaths of children were attributed to the

---

[40] U.S. Department of Health and Human Services, Centers for Disease Control and Prevention. "Alcohol Related Traffic Fatalities involving Children–United States, 1996", *Morbidity and Mortality Weekly Report* 1997; 46(48): 1130-1131.
[41] P.E. Bijur, M. Kurzon, *et al.*, "Parental Alcohol Use, Problem Drinking, and Children's Injuries", *Journal of the American Medical Association*. 267(23): 3166-71 June 1992. Serious injuries were those resulting in hospitalization, surgical treatment, missed school, and/or one half day or more spent in bed.
[42] S.W. Marshall, C.W. Runyan, *et al.*, "Fatal Residential Fires: Who Dies and Who Survives?" *Journal of the American Medical Association*. 279(20): 1633-1637, May 1998.

parents failure to perceive and respond to a fire emergency because of impairment of their sensory, judgment, or physical functions by alcohol consumption."[43]

A small child often will hide behind a bed or in a corner during a fire while awaiting help from parents or caretakers. In the case of an alcohol-impaired caretaker, the help is unlikely to arrive and the child is left behind. Furthermore, a rapidly evolving fire leaves little chance for a firefighter to find and rescue the child in time.

Elderly and disabled individuals who need help are also at high risk for death from fires, much like young children. An Assistant Fire Marshal of the Memphis Fire Department stated that many elderly individuals had died in fires because their caretakers, often their own children, were too impaired by alcohol to recognize the fire and render assistance in time.[44]

Gender Effects

There are wide gaps between male and female fire death rates, unintentional injury rates, and drinking rates in the United States. There is a wealth of data documenting this. Males comprise the majority of fatal and non-fatal injuries of all types year after year. They also have higher drinking rates and greater total alcohol consumption.

Fire injuries and deaths are not exempt from this pattern. Men continue to have almost twice as many fire deaths as women among nearly all age groups.[45] Over the past two decades, data from the NFIRS has consistently found the ratio of male to female fire deaths to be between 1.6 and 2.0 to 1. Additional evidence from the medical profession supporting this finding comes from the landmark study performed by Berl *et al.*, from the Johns Hopkins University. In it, male fire fatalities in the State of Maryland were shown to 50 percent higher than female fire fatalities.[46]

---

[43] W.S. Hollis, *et al.*, "Drinking: Its Part in Fire Deaths," *Fire Journal*, May 1973.

[44] W.S. Hollis, *et al.*, "Drinking: Its Part in Fire Deaths," *Fire Journal*, May 1973

[45] U.S. Fire Administration, *Fire in the United States: 1987-1996,* Eleventh Edition, FA-173/August 1999, (Emmitsburg, MD: U.S. Fire Administration).

[46] W.G. Berl and B.M. Halpin, "Human Fatalities From Unwanted Fires," Applied Physics Laboratory, The Johns Hopkins University. FPP TR 37, December 1978.

Data from the NFIRS found that men outnumber women by between 1.5 to 1.8 to 1 for fire injuries as well.[47] Medical data again concur with this finding. While conducting an injury survey among patients in four emergency room/trauma centers, researchers found that men outnumbered women in all six of the most commonly seen injuries. The cause of injury with the broadest contrast between genders was fire. Male fire injuries outnumbered female fire injuries in this study by about 3½ times.[48]

This pattern of gender differences in injury rate is not unique to burn injuries. Males are injured more often than females in almost all mechanisms of injury. The male dominant pattern is especially pronounced in the adolescent and young adult. This may indicate that males engage in more high risk behaviors (including alcohol abuse) than females.[49]

Alcohol and drinking patterns vary between genders and may help explain the differing casualty rates. National estimates show that males tend to consume larger quantities of alcohol with greater frequency than females. Men also vastly outnumber women when identifying problematic drinking behaviors, such as binge drinking. In the CDC survey on alcohol consumption, 62 percent of the male respondents stated that they drank an alcoholic beverage at least once a month. Only 44 percent of the female respondents consumed alcohol at least once a month. As the number of days of alcohol consumption increased, a greater percentage of men than women responded affirmatively. In the higher frequency brackets males outnumbered females in number of days spent drinking per month by about 2 to 1. Females outnumbered men (1.3 to 1) only in the 1 to 5 days per month category. When asked how many alcoholic beverages were consumed per drinking session, the number of men claiming 5 drinks or more per occasion was nearly 3½ times that of women. This response suggests that men in the United States drink more intensely and more frequently than their female counterparts.[50]

---

[47] U.S. Fire Administration, *Fire in the United States: 1987-1996,* Eleventh Edition, FA-173/August 1999, (Emmitsburg, MD: U.S. Fire Administration).

[48] C.J. Cherpitel, "Injury and the Role of Alcohol: Countywide Emergency Room Data." *Alcoholism, Clinical & Experimental Research.* 18(3):679-84, June 1994.

[49] N.S. Redeker, S.C. Smeltzer, J. Kirkpatrick and S. Parchment. "Risk Factors of Adolescent and Young Adult Trauma Victims." *American Journal of Critical Care.* 1995. Sep;4(5):370-8.

[50] Behavioral Risk Factor Surveillance System, "Nationwide Alcohol Consumption," Online Prevalence Data, Centers for Disease Control and Prevention, 1995

It has been theorized that the drinking behavior commonly seen in the male population is directly related to their high risk for injuries and fatalities. There is reason to believe that a greater number of fire fatalities that were under the influence of alcohol are male as well. Berl *et al.* reported that of fire fatalities examined in Maryland, males comprised 72 percent of those with BALs over 0.1 percent.[51] The state of Minnesota, one of the only states requiring BALs for all fire deaths, found similar results. For the years 1993 to 1996, approximately 80 percent of fatalities with positive BALs were men. The results of a Minnesota study will be discussed in a separate section.

Leading Causes of Fire Casualties

Smoking

Nationally, fire-reporting agencies have identified smoking as the fifth most frequent cause of residential fires, the leading cause of fire fatalities, and the second most common cause of fire-related injuries.[52] Smoking combined with alcohol use creates an even greater risk for fire injuries and fatalities, as evidence suggests that the two exert a synergistic effect on each other.

Studies have also shown that smokers are more inclined to engage in risk-taking behavior and are more commonly involved in accidents. Researchers have concluded that binge drinking (as well as other specific risk-related behaviors) was correlated to smoking among the youth in the United States.[53] Many of the various risk factors for injuries in general, including fire injuries, are inter-related and magnify the effects of the other.

A relatively new phenomenon of "social drinking" has been identified by researchers that may, in turn, lead to alcohol-related fire fatalities. Social smoking is a trend identified in the college age population where the individual smokes only when in specific social situations. These situations include parties and bars.[54] Additional research

---

[51] W.G. Berl and B.M. Halpin, "Human Fatalities From Unwanted Fires," Applied Physics Laboratory, The Johns Hopkins University. FPP TR 37, December 1978.

[52] U.S. Fire Administration, *Fire in the United States: 1987-1996,* Eleventh Edition, FA-173/August 1999, (Emmitsburg, MD: U.S. Fire Administration).

[53] L.G. Escobedo, "Relationship Between Cigarette Smoking and Health Risk and Problem Behaviors Among U.S. Adolescents, " *Archives of Pediatric & Adolescent Medicine*. 151(1):66-71, January 1997.

[54] D. Hines, A.C. Fretz and N.L. Nollen, "Regular and Occasional Smoking by College Students: Personality Attributions of Smokers and Non-Smokers." *Psychological Reports*. 1998. Dec;83(3 pt 2):1299-306.

has linked social smoking to binge drinking and severe alcohol intoxication.[55] The link between smoking and fire has been well established through fire data analyses. The effects of alcohol consumption would likely lead to an inattentiveness to lit cigarettes and, potentially, to fire. Social smoking may become a significant link between alcohol abuse and fire.

Researchers from the University of Chicago Department of Psychiatry concluded that smokers consume more alcohol than do non-smokers, heavy drinking tends to be associated with heavy smoking, and a large majority of alcoholics, characterized by heavy drinking, are smokers.[56] Another study reviewing people aged 40 to 49 years who received check-ups from 1979 to 1985, found alcohol use to be strongly associated with the number of cigarettes smoked per day.[57] Alcohol and smoking represent independent fire safety risk factors. Alcohol intoxication leaves the drinker bereft of control and mental acuity. Smoking has been associated with the ignition of many fires in which the smoker is intimately involved, a factor that significantly contributes to the severity of smoking fire-related injuries and deaths.[58] When used together, alcohol and smoking increase one's chance of starting a fire while at the same time decrease the chances of detecting, mitigating, and escaping the fire.

Interviews from households in King County, Washington that had sustained a fire-related injury or fatality were conducted in an effort to identify the risk of fire injury in relation to the incidence of smoking and alcohol use. Using a control group of households that had not experienced a fire emergency, researchers concluded that households with alcohol drinkers who consumed 6 or more drinks per occasion increased their risk for burn injury by 8 times.[59] Further analysis revealed that households with higher drinking levels were associated with higher smoking levels as well. Heavy or

---

[55] J.B. Schorling, M. Gutgesell, P. Klas, D. Smith and A. Keller. "Tobacco, Alcohol and Other Drug Use Among College Students." *Journal of Substance Abuse*. 1994. 6(1):105-15.
[56] J.P. Zacny, "Behavioral Aspects of Alcohol-Tobacco Interactions," *Recent Developments in Alcoholism*. 8:205-219, 1990.
[57] G.D. Friedman, I. Tekawa and A.L. Klatsky, "Alcohol Drinking and Cigarette Smoking: An Exploration of the Association in Middle-Aged Men and Women," *Drug and Alcohol Dependence*. 27(3):283-290, May 1991.
[58] U.S. Fire Administration, National Fire Incident Reporting System
[59] J.E. Ballard, T.D. Koepsell and F. Rivara, "Association of Smoking and Drinking with Residential Fire Injuries," *American Journal of Epidemiology*. 135 (1):26-34, January 1992.

binge drinking was identified as a predictor for fires and burn injuries, and when coupled with smoking, appeared to be even more of a significant factor.[60]

Arson

Arson is the second leading cause of fire deaths in the United States. Alcohol may be a factor contributing to the act of arson and hence, to fire casualties. Alcohol has long been associated with violent crimes in this country and around the world. Arson, especially when associated with residential fires, is a violent crime meant to hurt, kill, and destroy. A variety of motives can be attributed to arson acts, but alcohol is most prevalent in incidents of arson that are meant to inflict physical harm.[61]

Social scientists have theorized as to why alcohol seems to promulgate criminal activity. Some theories suggest that alcohol could be directly linked to crime through its pharmacological properties or indirectly through social learning about typical behavior under the influence. MacAndrew and Edgerton (1969) showed that "at least some part of drunken comportment is learned."[62] They also showed that an important factor in drunken behavior is that in some cultures, including our own, alcohol can serve to initiate a normative "time out" where people are allowed to deviate from social norms.[63] While this research is 30 years old, there is little information readily available to indicate that it is invalid today.

Researchers have also discovered a strong association between heavy drinking and criminal behavior, both in terms of the need to engage in criminal behavior and in terms of perceptions of the risk and moral condemnation associated with it. MacAndrew and Edgerton found that heavy drinking led to unstable moral evaluations and irrational perceptions regarding the consequences of their behaviors.[64] Such instability in moral

---

[60] J.E. Ballard, T.D. Koepsell and F. Rivara, "Association of Smoking and Drinking with Residential Fire Injuries," *American Journal of Epidemiology*. 135 (1):26-34, January 1992.

[61] Ibid.

[62] MacAndrew and Edgerton (1969), as cited in L. Lanza-Kaduce, D. M. Bishop, L. Winner, "Risk Benefit Calculations, Moral Evaluations, and Alcohol Use: Exploring the Alcohol-Crime Connection," *Crime and Delinquency*. Vol. 43, no.2, 1997.

[63] Ibid.

[64] L. Lanza-Kaduce, D.M. Bishop, L. Winner, "Risk/Benefit Calculations, Moral Evaluations, and Alcohol Use: Exploring the Alcohol-Crime Connection." *Crime and Delinquency.* Vol. 43, no. 2, 1997.

evaluations was not identified among moderate or light drinkers or abstainers, even in the same social situations.[65]

A recent study in Minnesota that examined the relation of drugs and alcohol to arson found a prominent correlation between alcohol and arson fires. Since 1994, the state of Minnesota has required autopsies to be performed on all fire-related deaths, one of the only states to do so. The data show that nearly 48 percent of all Minnesota fire deaths involve drug and/or alcohol use. The state then became interested in learning the extent to which drugs and alcohol played a part in incendiary fires so that they could develop strategies to minimize the number of any future arson fires with those contributing factors.[66]

The study found evidence of an arson trend in which socio-economic deprivation, coupled with alcohol use resulted in the commission of arson. Similarly, there was evidence of alcohol use among suspects who set fires to fraudulently collect on insurance policies. In these cases, alcohol may have been used to give suspects the "courage" to commit the act, to dull their sense of wrongdoing, or to help them rationalize their own acts since they were drunk at the time.[67]

International Studies

Similar patterns of alcohol usage that correlate to fire are found in other cultures. The use of alcohol appears to be a significant factor in explaining international differences in fire death rates. Approximately 80 percent of Finnish male fire fatalities were found to have been drinking at the time of their death.[68] The association between house fire fatalities and high blood alcohol concentrations has also been recognized as a particular problem in Scotland.[69] The death rate for fatal fire accidents in Denmark has

---

[65] Ibid.

[66] United States Fire Administration, Minnesota Department of Public Safety, Office of the State Fire Marshal, and TriData Corporation, *The Connection Among Drugs, Alcohol, and Arson in Minnesota,* April 1998.

[67] Ibid.

[68] R. Hohanen, *et al.,* "Males as a High-Risk Group for Trauma," *The Journal of Trauma.* 30:155-162, 1990.

[69] T. Squires, A. Busuttil, "Alcohol and House Fire Fatalities in Scotland, 1980-1990," *Medicine, Science & the Law.* 37(4):321-5, 1997.

doubled since 1951, attributed mostly to fire accidents associated with smoking, which often included alcohol intoxication.[70]

---

[70] P. Leth, M. Gregersen, S. Sabroe, "Fatal Residential Fire Accidents in the Municipality of Copenhagen, 1991-1996." *Preventive Medicine*.27(3):444-51, May-June 1998.

VII. SMOKE ALARMS AND THE ALCOHOL-IMPAIRED FIRE CASUALTY

Attempting to quantify the extent to which alcohol impairs one's ability to detect and escape from a fire is difficult. In an effort to establish a clear relationship between the ability to perceive a fire and subsequently escape from it, researchers have examined the relationship between alcohol impairment and the function of smoke alarms. The presence of a working smoke alarm significantly reduces the risk of becoming a fire fatality.[71, 72] Many fire prevention programs revolve around installing and maintaining smoke alarms. However, existing evidence seems to point to alcohol impairment as being a strong independent risk factor for deaths caused by fires. In a study performed by Runyan *et al.*, the fire fatality risk ratio for an alcohol-impaired person was more than double the risk ratio of a person living without a smoke alarm.[73]

Smoke alarms have been found to be most protective against death from a fire for small children, but only when alcohol and/or drugs do not impair the caretaker.[74] Children who are unable to help themselves depend upon their parents for escaping a fire. Failure to respond to a smoke alarm in adequate time allows a fire to rapidly evolve and significantly reduce ones chances of escaping. Even if an impaired parent survives, it is often too late to retrieve the child, or escape together.

---

[71] L.E. Smith, *Fire Incident Study: National Smoke Detector Project*. EPHA, Directorate for Epidemiology, US Consumer Product Safety Commission. Bethesda, MD. January 1995; page 3.

[72] As cited in "Perspectives in Disease Prevention and Health Promotion Progress Toward Achieving the National 1990 Objective for Injury Prevention and Control," *Morbidity and Mortality Weekly Report*. 37(9): 138-149, March 1988; E. McLoughlin *et al.* "Smoke Detector Legislation: Its Effect on Owner-occupied Homes," *American Journal of Public Health*. 75: 858-862, 1985.

[73] C.W. Runyan, *et al*, "Risk Factors For Fatal Residential Fires," *New England Journal of Medicine*.327(12):859-863, 1992.

[74] Ibid.

VIII. CASE STUDY: RESULTS FROM THE ONTARIO FIRE REPORTING SYSTEM

Background

Ontario, one of Canada's largest provinces, comprises over one third of Canada's population. In 1997, Ontario experienced a total of 139 fatal fires, which killed 154 persons.[75] The Ontario Fire Reporting System requires all fire departments to report and record every fire incident and casualty. This information is then registered with the local provincial governments. A unique and helpful attribute of OFRS is that separate categories which distinguish alcohol impairment from impairments caused by other drugs are used to describe the condition of the fire victim prior to injury or death.

Fire Casualties

From 1990 to 1995, Ontario experienced a total of 5,934 casualties from fire. [Note: 1993 has been omitted from this study due to reporting errors. As a consequence, the time frame reflects five years rather than six.] Of these casualties, 410 were impaired by alcohol at the time of their injuries and 7 were impaired by drugs. Alcohol-impaired casualties resulted in 178 fatalities and 232 injuries (Table 1). While alcohol-impaired casualties only accounted for 7 percent of all Ontario's fire casualties, alcohol-impaired victims accounted for 22 percent of all fire deaths (Table 2).

Table 1. Distribution of Ontario Fire Casualties

Type of Casualty	Total	Deaths	Percent of Total	Injuries	Percent of Total
All Fire Casualties	5,934	797	13%	5,137	87%
Non-Alcohol-Impaired Casualties	5,524	619	11%	4,905	89%
Alcohol-Impaired Casualties	410	178	43%	232	57%

---

[75] Data for this section comes from the Office of the Fire Marshall of Ontario

Table 2: Alcohol-Impaired Casualties' Prominence in Ontario Fire Deaths

	All Fire Casualties	Alcohol-Impaired Casualties	Percent of All Fire Casualties
Total Casualties	5,934	410	7%
Deaths	797	178	22%
Injuries	5137	232	5%

The proportion of fire fatalities was markedly different between the alcohol-impaired and non-alcohol-impaired casualties. Alcohol-impaired victims accounted for a larger proportion of fire fatalities than injuries. Fire victims who were not under the influence of alcohol were far less likely to killed (11 percent) than to be injured. In the alcohol-impaired group, fatalities represented 43 percent of the casualty total. The sharp rise in the percentage fatalities when alcohol was involved suggests that alcohol affects the severity of a fire casualty. Alcohol-impaired fatalities occurred nearly four times as often as non-alcohol-impaired fatalities and over three times as often as fire fatalities in general. Alcohol impairment appears to lead to more serious, fatal, injuries from fires.

Demographics of Fire Casualties

Age

One-quarter of the alcohol-impaired fire fatalities were aged 25 to 34, despite the fact that the same age group accounted for only 16 percent of fire deaths overall and 14 percent of the non-alcohol-impaired casualties (Figure 1). Victims aged 35 to 44 and 45 to 54 followed at 19 percent and 16 percent respectively.

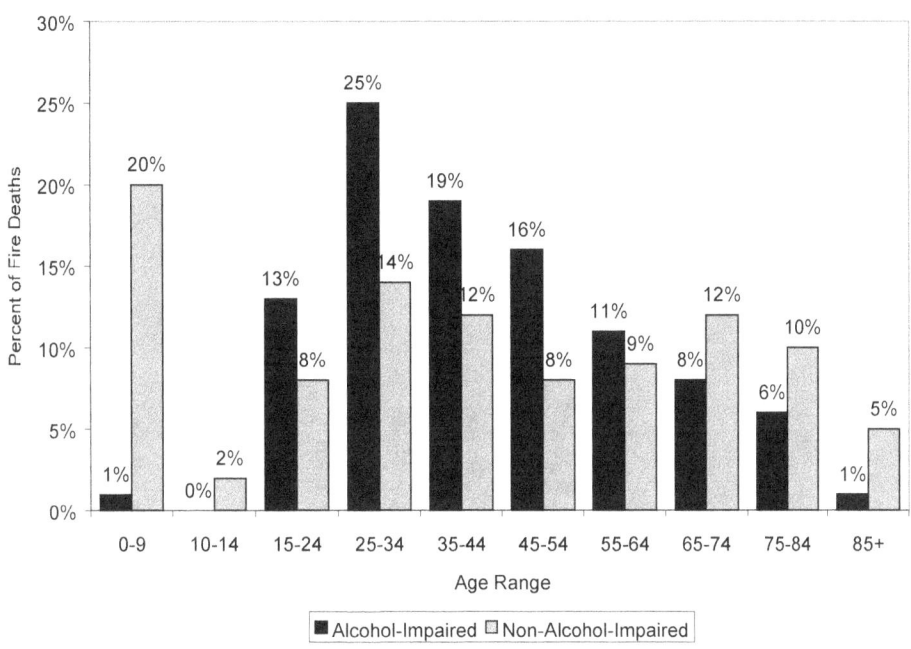

Figure 1: Alcohol-Impaired vs. Non-Alcohol-Impaired Fire Deaths by Age (Ontario 1990-1995, excluding 1993)

Approximately 35 percent of all fire deaths aged 25 to 34 and aged 45 to 54 (45 of 129 fatalities and 28 of 79 fatalities, respectively) were impaired by alcohol; 15 to 24 year-olds and 35 to 44 year-olds exhibited similar percentages at 34 percent and 31 percent, respectively. These four consecutive age groups (ages 15 to 54) accounted for not only 75 percent of alcohol-impaired deaths, but also a large proportion (one-third) of fire deaths in their age groups. By comparison, the 15 to 54 age groups accounted for only 42 percent of non-alcohol-impaired deaths (Figure 2).

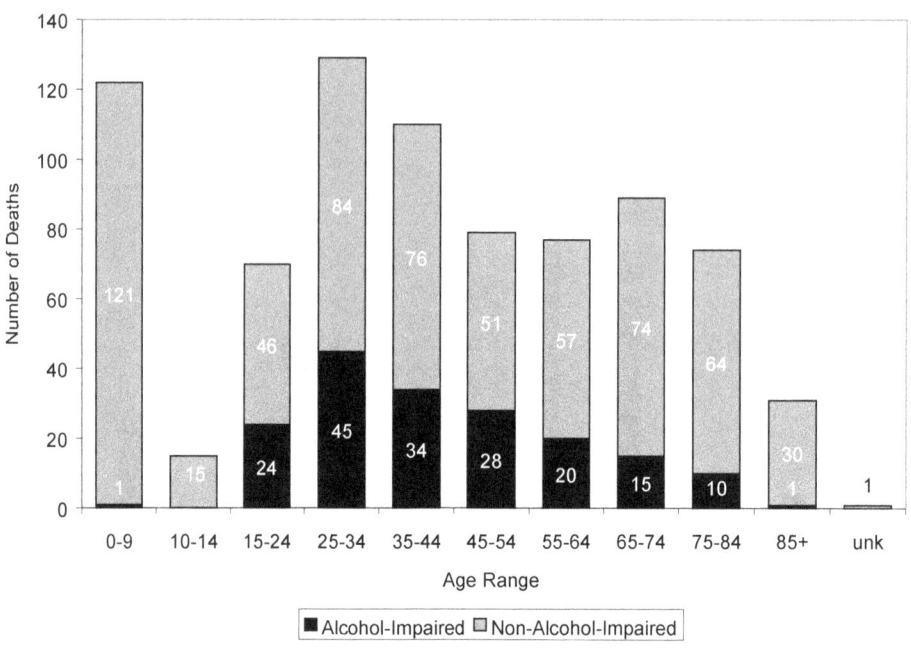

Figure 2: Distribution of Alcohol-Impaired vs. Non-Alcohol-Impaired Fire Deaths (Ontario 1990-1995, excluding 1993)

For those sustaining injuries (Figure 3), 35 to 44 year-olds comprised the largest portion of alcohol-impaired fire injuries (29 percent), followed by 25 to 34 year olds (27 percent). Together, these two age groups accounted for over half (56 percent) of all fire-related injuries in which the individual was impaired by alcohol.

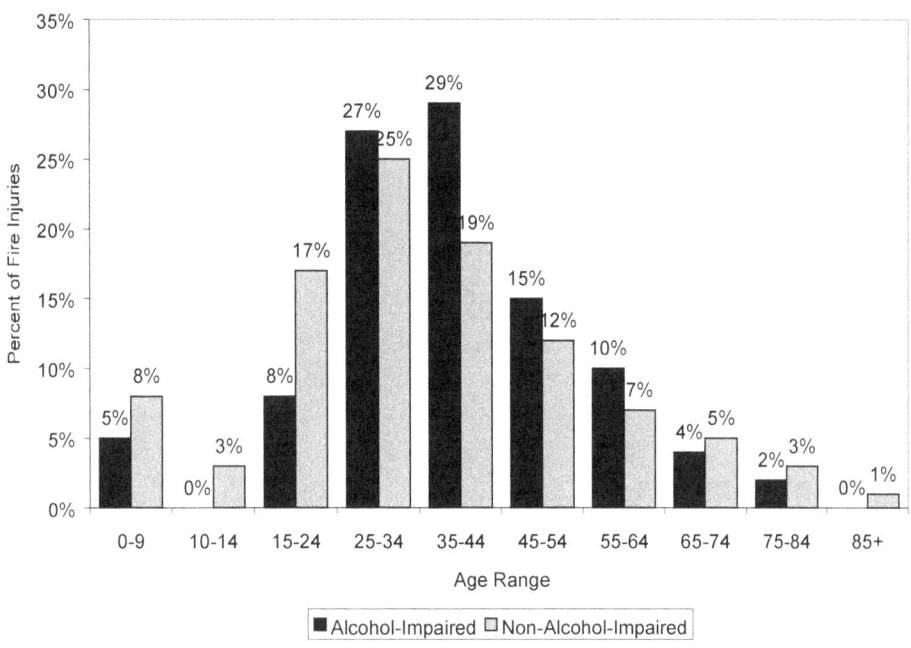

Figure 3: Alcohol-Impaired vs. Non-Alcohol-Impaired Fire Injuries by Age (Ontario 1990-1995, excluding 1993)

Gender

From 1990 to 1995, men had twice as many fire deaths as women; male fire fatalities accounted for two-thirds (66 percent) of all fire fatalities. This ratio rose, as did the disparity, between male and female fatalities when alcohol was involved. Among the alcohol-impaired fatalities, men outnumbered women by nearly 3 to 1, accounting for 75 percent of alcohol-impaired fatalities (Table 3).

Table 3: Male to Female Fire Casualty Ratios in Ontario

	Deaths	Injuries
All Casualties	2.0	1.8
Alcohol-Impaired	2.8	4.8

A similar trend was found for all fire and alcohol-impaired fire injuries (Table 3). Men accounted for 64 percent of all fire-related injuries in general and 83 percent of those involving alcohol. Although men already outnumbered women nearly 2 to 1 in all fire-related injuries, this ratio rose to nearly 5 to 1 when the injured parties were impaired by alcohol.

29

When examining male and female drinking patterns in Ontario, these discrepancies in fire casualties might appear puzzling. According to the Toronto Addiction Research Foundation, approximately equal proportions of the male and female population over the age of 18 were current drinkers – 85 and 80 percent respectively. However, when the same group was asked if they consumed more than 5 drinks in a single sitting, 13 percent of males and 4 percent of females responded yes. In addition, 7 percent of males and 3 percent of females claimed to be daily drinkers. In effect, heavy drinkers were more than three times as likely to be men. Daily drinkers were more than twice as likely to be women.

Causes of Fire Casualties

Smoking (53 percent) followed by cooking (23 percent) and open flame fires (14 percent) were the three leading causes for alcohol-impaired fatalities (Figure 4). For non-alcohol-impaired fire deaths, the leading cause was smoking (35 percent), followed by open flame (28 percent), and cooking (16 percent). More alcohol-impaired fires occurred as a result of smoking fires than non-alcohol-impaired victims deaths (53 percent vs. 30 percent).

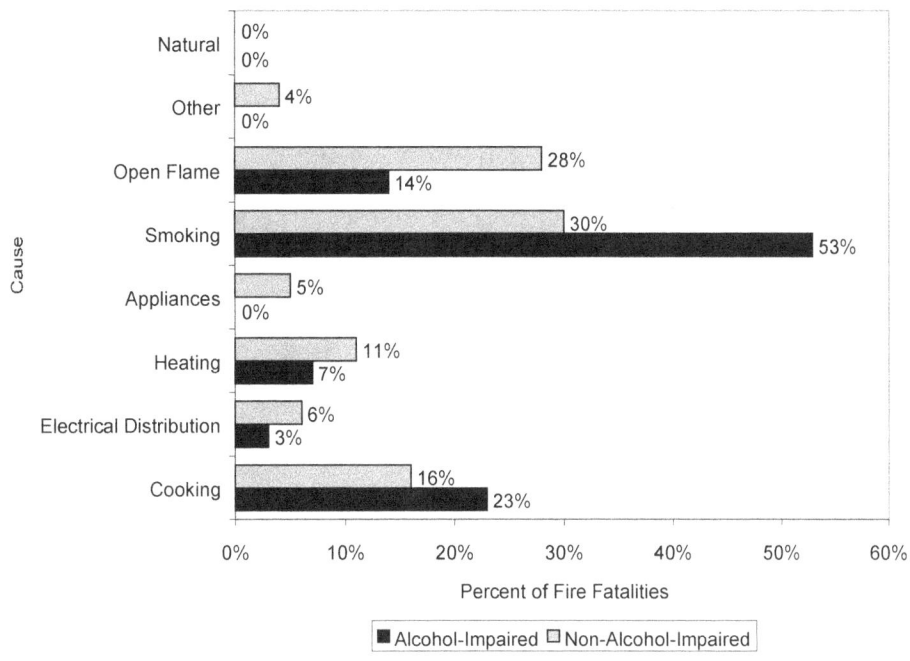

Figure 4: Alcohol-Impaired vs. Non-Alcohol-Impaired Fire Fatalities by Cause (Ontario 1990-1995, excluding 1993)

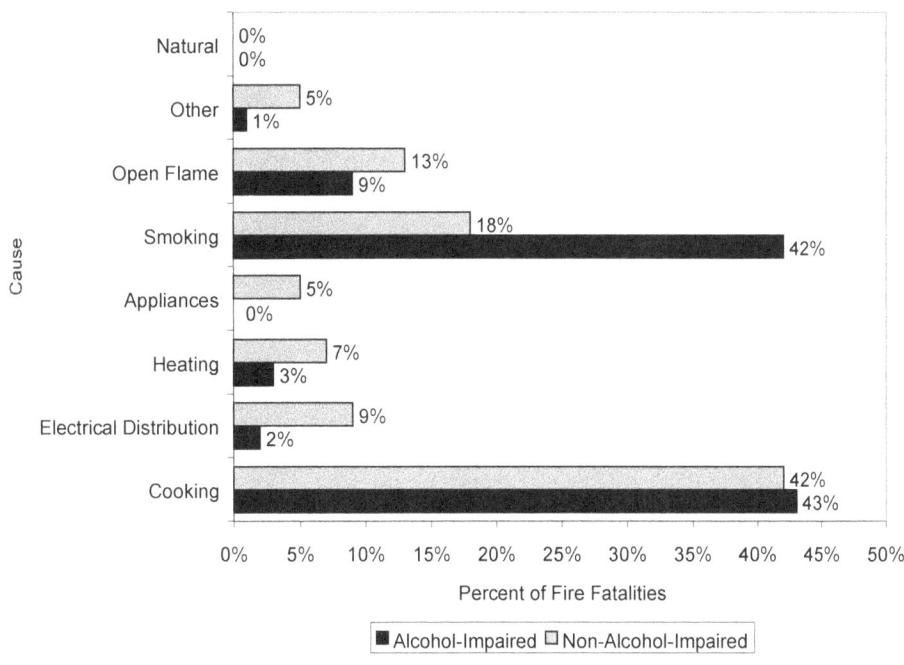

Figure 5: Alcohol-Impaired vs. Non-Alcohol-Impaired Fire Injuries by Cause (Ontario 1990-1995, excluding 1993)

31

The three leading causes for alcohol-related fire injuries were cooking (43 percent), smoking (42 percent), and open flame (9 percent) (Figure 5). Among non-alcohol-impaired injuries, cooking fires (42 percent) were also credited with causing the most injuries, followed again by smoking fires (18 percent). As for fire deaths, more alcohol-impaired fire injuries were injured in a smoking fire than non-alcohol-impaired victims (42 percent vs.18 percent). Although cooking fires were the leading cause of injuries for both alcohol-impaired and non-alcohol-impaired persons, the percentage of cooking-related injuries differed by only 1 point between the two groups.

Alcohol Consumption and Drinking Patterns

The legal drinking age in Ontario during this period was 19. Alcohol consumption in Ontario comes with a relatively high degree of social acceptance. It is not illegal for parents to give alcohol to a child under the legal drinking age at home. In a 1990 nation-wide Gallup poll for Canada, 79 percent of adults over the age of 15 reported that they had at some point consumed alcohol. Over half the adult population (55 percent) said that they have 6 drinks or more in a sitting and 10 percent reported using alcohol daily. Among young people aged 12 to 19 years, 73 percent stated that they had consumed alcohol at least once in the previous year. More than 1 in 5 of this age group admitted to drinking more than once a week. The United States and Canada have very similar alcohol consumption rates per capita; in both the United States[76] and Canada[77], approximately 2.6 gallons of alcohol are consumed per year by people aged 15 and over.

Unintentional Injury Deaths

Accidents and adverse effects are the fifth leading cause of death in Canada[78], as in the United States.[79] Unintentional injury death rates are approximately 30 deaths per 100,000 people in both the United States[80] and Canada.[81] In both countries, the age 15 to

---

[76] Centers for Disease Control and Prevention. "Apparent Per Capita Ethanol Consumption – U.S., 1977-1986," *Morbidity and Mortality Weekly Report* 1989. Nov 38 (46); 800-803.
[77] Statistics Canada. "Health Indicators," National Population Health Survey. 1996. Electronic media. Catalog no. 82-221-XDE
[78] Statistics Canada. "Causes of Death," 1996 electronic media. Catalog no. 84-208-XP8
[79] U.S. Department of Commerce, Bureau of the Census. *The Statistical Abstract of the United States – 1993 113th ed.* 1993. Washington, DC:GPO.
[80] U.S. Department of Commerce, Bureau of the Census. *The Statistical Abstract of the United States – 1993 113th ed.* 1993. Washington, DC:GPO.

34 population is the most active age bracket for drinking and experiences the highest unintentional injury death rate. Motor vehicle crashes are the leading cause of unintentional injury deaths involving alcohol in both countries as well. While less studied, the results from the Ontario Fire Reporting System supports the possibility that fire casualties are prominently affected by alcohol as well.

[81] Statistics Canada. "Causes of Death," 1996 electronic media. Catalog no. 84-208-XP8

IX. CASE STUDY: FIRE FATALITIES IN MINNESOTA

Background

In 1991, the Minnesota Center for Health found that unintentional injury fatalities accounted for 31 percent of all alcohol-related deaths in the state.[82] This was by far the majority of all fatalities involving alcohol. Of these deaths, 6 percent stemmed from fires.[83] In an effort to learn more about the nature and extent of the alcohol problem as it is related to injuries, the Minnesota Department of Public Safety began recording BALs for all unintentional injury deaths, including fire deaths. It is this practice that makes examining fire deaths in the state of Minnesota a useful tool by which to study the impact of alcohol on fire fatalities.

Fire Fatalities

There were a total of 255 fire fatalities in the state of Minnesota from 1993 to 1996.[84] Of these, 74 were found to have positive blood alcohol concentrations, representing 30 percent of all fire fatalities and 40 percent of all fire deaths over the age of 15. It is important to note that autopsy reports were the source of the data and positive blood alcohol results of any measurement were recorded, not necessarily the legal intoxication standard.

Ninety percent of alcohol-impaired fatalities met or exceeded the legal intoxication standard 0.1 percent BAL (Figure 6). The BALs for over half of the fatalities positive for alcohol were between 2 and 4 times over this limit. Most of the alcohol-impaired fire fatalities were severely intoxicated, often to the point of stupor.

---

[82] D. Kloehn, K. Miner, K. Daly, *Alcohol Use in Minnesota: Extent and Cost*, Minnesota Department of Health, October 1995.
[83] Ibid.
[84] Minnesota Office of the State Fire Marshal.

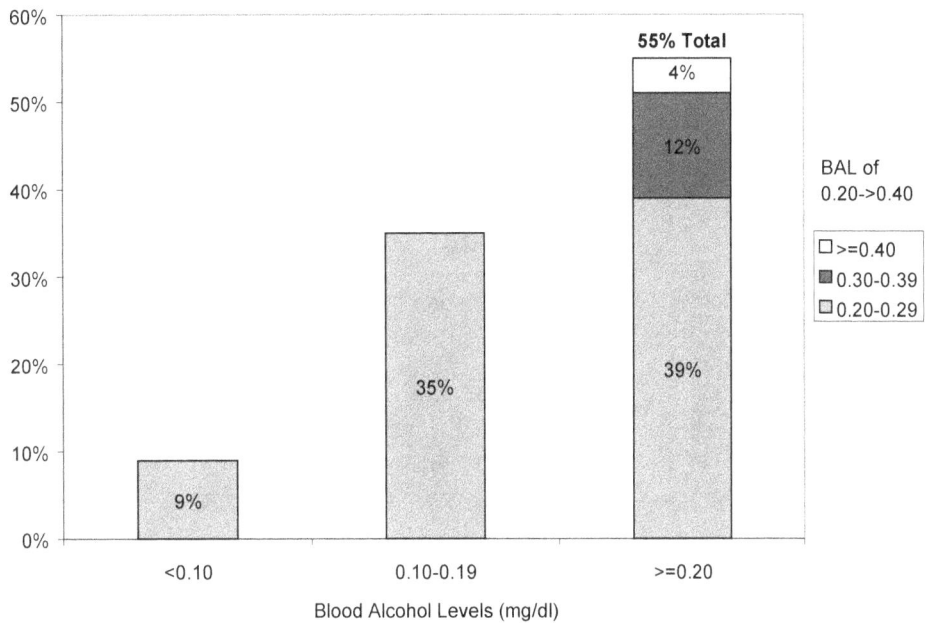

Figure 6: Minnesota Fire Fatalities with Positive Blood Alcohol Levels by Blood Alcohol Level (1993-1996)

Demographics of Fire Fatalities

Age

Minnesota fire victims aged 25 to 34 accounted for the largest percentage (27 percent) of fire deaths that were under the influence of alcohol at the time of their death. The age groups 35 to 44 and 15 to 24 followed closely behind and comprised 23 percent and 21 percent of all alcohol-impaired fire fatalities, respectively. Combined, these three age brackets represented approximately 70 percent of all Minnesota fire fatalities with measurable BALs (Figure 7).

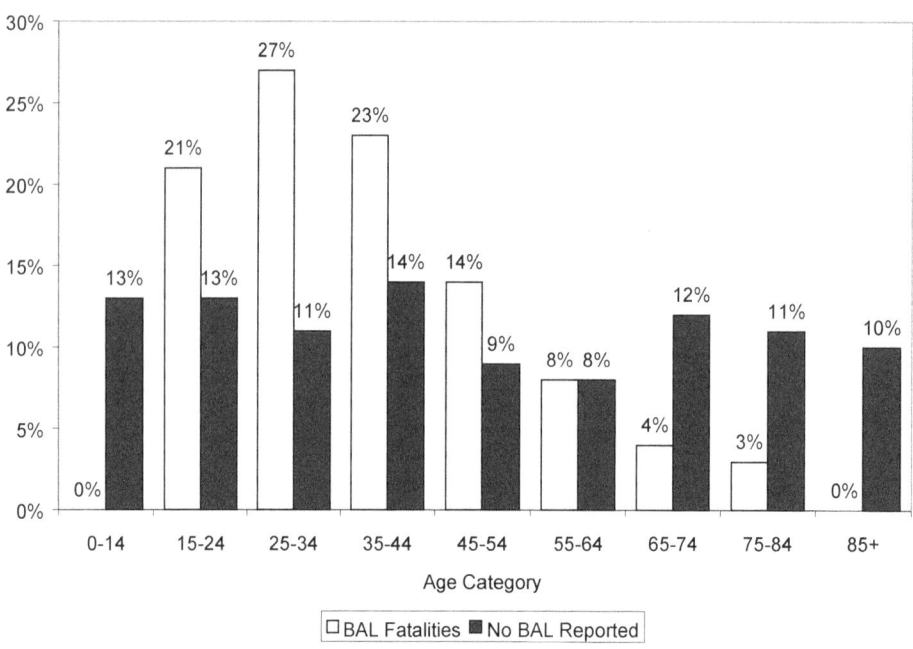

Figure 7: Minnesota Fatalities by Age Group (1993-1996)

Approximately 40 percent of fire fatalities over the age of 15 tested positive for alcohol use at the time of their death. Fire death victims aged 25 to 34 comprised the largest number of alcohol-impaired fire deaths, and at the same time represented the largest percentage of fire fatalities with respect to their age group; approximately 63 percent (19 of 40 fatalities) of all Minnesota deaths aged 25 to 34 were under the influence of alcohol. The age groups 35 to 44 and 45 to 54 followed closely behind; alcohol-impaired fatalities accounted for 53 percent of all fire deaths in each age group. Alcohol-impaired fatalities aged 15 to 24 also accounted for 43 percent of fire deaths in their age bracket. The 15 to 54 age group accounted for 82 percent of alcohol-impaired casualties but only 31 percent of non-alcohol-impaired deaths (Figure 8).

36

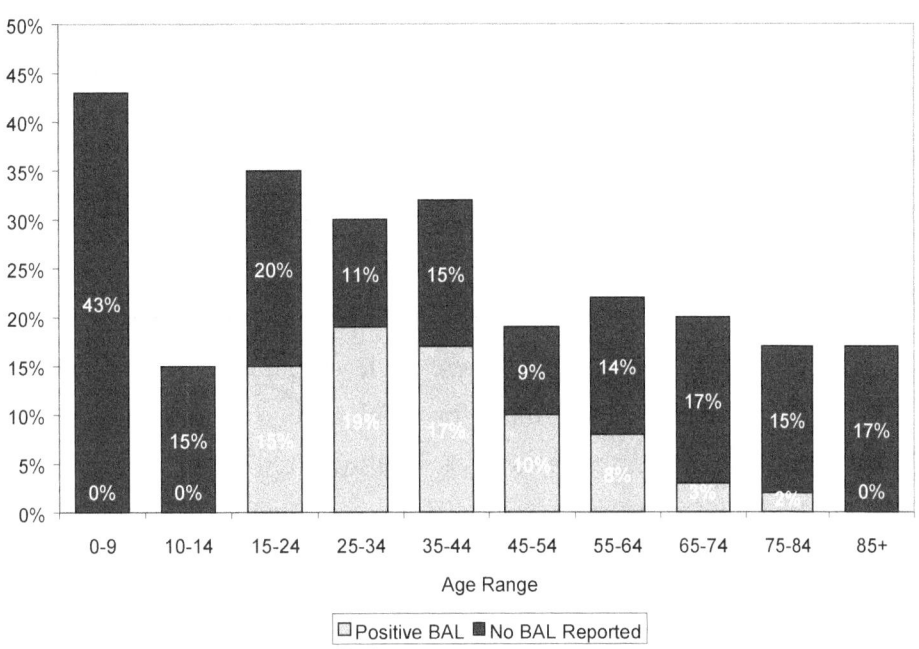

Figure 8: Distribution of Minnesota Fire Fatalities by Age, 1993-1996

Gender

Men represented 65 percent of total fire fatalities and 80 percent of fire fatalities with positive BALs in Minnesota. The male to female ratio for fire deaths in general was nearly 2 to 1, but when alcohol was involved, the ratio rose to 4 to 1 (Table 4). For those fatalities with BALs meeting or exceeding the legal intoxication standard of 0.1 percent BAL, 87 percent were men. In addition, alcohol-impaired fire fatalities with BALs equal to or greater than 0.4 percent were comprised entirely of men. Conversely, women outnumbered men for fatalities that had BALs below the intoxication standard, accounting for 74 percent of deaths with BALs less than 0.1 percent.

Table 4: Male to Female Fire Death Ratios in Minnesota

Type of Casualty	Ratios
All Casualties	1.9
Positive BALs	2.8

This pattern is reflected throughout all alcohol-related deaths in Minnesota, regardless of the cause. According to the Minnesota Department of Health, significant gender differences among alcohol-related deaths can be explained by the exaggerated drinking patterns of males. According to the Minnesota Department of Health, 28 percent of male adults report binge drinking as opposed to only 10 percent of adult females.[85]

Fire Fatality Causes

Smoking- and heating-related fires were the leading causes of fire deaths among victims with positive BALs as well as those for which no BAL was recorded (Figure 9). Although the percentages of deaths between alcohol-impaired and non-alcohol-impaired deaths were similar for heating fires, a stark difference exists for smoking fires. Nearly two-thirds (64 percent) of the alcohol-impaired fire deaths were killed in a smoking fire, as opposed to 37 percent of the non-alcohol-impaired fatalities. Approximately 46 percent of deaths caused by smoking fires had positive BALs. In addition, over one-third of cooking fire fatalities and one-fourth of heating fire fatalities were under the influence of alcohol at the time of their death.

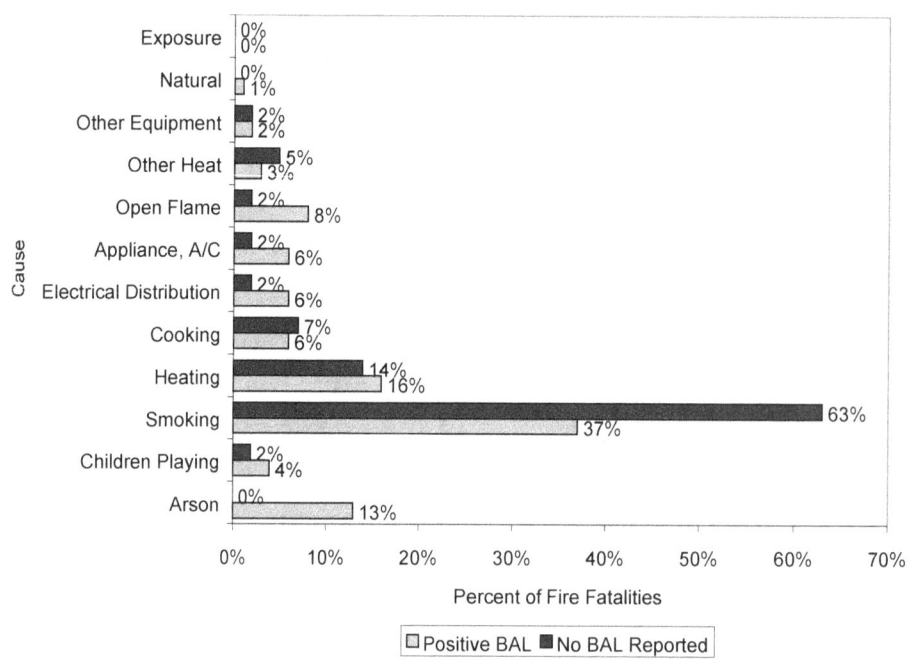

---

[85] Ibid.

Figure 9: Minnesota Fire Fatalities by Cause, 1993-1996

Alcohol Consumption and Drinking Patterns

Per capita, alcohol use in Minnesota exceeds that of use nationwide. In 1991, the average drinker in Minnesota consumed an average of more than 2 drinks per day. Approximately one-third of high-school seniors in Minnesota drink to the point of intoxication at least once a month, or have 5 or more drinks per sitting.[86]

Binge drinking is more common among young persons aged 18 to 34. Males make up the majority of these drinkers as well. Among Minnesotan's aged 18 and older, men are three times more likely than women to report binge drinking. Heavy drinking, (those who consume 60 or more alcoholic beverages in a month) is also much more prevalent in males. In Minnesota, the prevalence of binge drinking remains higher, and the rate of abstinence from alcohol lower, than in many other states.

Unintentional Injury Deaths

In 1991, there were 35,241 deaths in Minnesota. Nearly 1,500 deaths were caused by unintentional and accidental injuries. Of these deaths, 31 percent were under the influence of alcohol. The top four causes of unintentional injury deaths in general were motor vehicle accidents, falls, fires and burns, and drownings. Motor vehicle crashes were responsible for the majority of alcohol-related deaths (54 percent), followed by falls (27 percent), fire-related injuries (6 percent), drownings (5 percent), and alcohol poisoning (3 percent).[87] According to the Minnesota Department of Health, for the population aged 35 and younger, almost every alcohol-related death (97 percent) was due to an alcohol-related injury or act of violence.[88]

---

[86] Ibid.
[87] Ibid.
[88] Ibid.

39

X. CONCLUSION

Human behavior exerts a powerful impact on the risk of fire. The comportment with which an individual acts not only affects the ignition of a fire, but how one manages and escapes from it. As alcohol can significantly affect human behavior, it is not surprising to find that it is a major factor in fire-related deaths and injuries. Alcohol intoxication has been identified as the strongest independent risk factor affecting residential fire deaths. By altering ones cognitive, physiological, and motor functions, alcohol increases the chance of starting a serious fire while at the same time reduces the chance of survival from a fire or burn injury.

The end result of acute alcohol intoxication is carelessness. It is this carelessness that generates the possibility of starting a fire. It reduces one's ability to properly detect fire and renders mechanical detection devices virtually useless. You cannot escape from a fire if you cannot sense it or heed a warning alarm. Chronic alcohol ingestion exhibits equally deleterious effects by weakening the body's defenses and healing mechanisms. Chronic alcoholic users have been shown to die from smaller burns and at a higher rate than non-drinking burn casualties.

Alcohol impacts a silent population as well. Children who are cared for by an alcohol-impaired adult sustain serious illness and injury at a much greater frequency than do children of non-alcohol-impaired adults. Through no fault of their own, children, senior citizens, and disabled individuals lose their lives in fires started by another's carelessness. Unable to help themselves, these people die while waiting to be rescued by someone who cannot even rescue himself or herself.

Smoking fires are the leading cause of fires that kill people. They leave the victim in a particularly vulnerable position because these individuals are typically intimately involved in the ignition of the fire. Often, the victim's clothes or nearby upholstery is the material ignited. Such intimacy significantly reduces one's ability to effectively mitigate or extinguish a rapidly evolving fire. Numerous studies have shown the correlation between alcohol consumption and smoking frequency. The facts strongly link increased drinking to increased smoking. Acting synergistically, both factors contribute to abnormally high risk for fire death and injury. Not only does alcohol impairment increase the propensity for starting a smoking fire; it increases the incidence and severity of potential casualties.

Drinking patterns seem to exert more of an effect on fire casualties than does total alcohol consumption. This impact has been demonstrated among all unintentional injuries, not only those from fire. Studies have examined the relationship between acute and chronic alcohol consumption and found that chronic users were more likely to sustain unintentional injuries than individuals who drank on an infrequent basis. In addition, alcohol abusers appear to suffer a disproportionately high number of fire fatalities and injuries relative to their percent of the total population. It seems likely that alcohol abuse may also be a major factor in explaining high fire death rates among nations with high alcoholism rates.

An ounce of prevention is worth a pound of cure. So the old saying goes and still holds true for fire prevention. Individuals who use and abuse alcohol are segments of the population with extremely high fire risks. While less studied than other accidental and unintentional injuries, fire casualties are emerging as an injury subset highly influenced by problematic drinking behaviors. Fire fatalities and injuries can be prevented if a concerted effort is made to identify and modify high-risk drinking patterns. It also may be possible to minimize fire risk by increasing the awareness of those who drink and those who are surrounded by regular drinkers. Successful educational campaigns have been launched warning people of the dangers of drunk driving and the same can be done to shed light upon the subtle dangers of alcohol and fire.

www.ingramcontent.com/pod-product-compliance
Lightning Source LLC
Chambersburg PA
CBHW081237170526
45165CB00009B/3083